I am NOT a DINOSAUR

Written and Illustrated by:

Kevin L. Brooks

I am NOT a DINOSAUR

Written and
Illustrated by:

Kevin L. Brooks

This is a work of nonfiction. The animals, places, and events are true.
However, some incidents either are the product of the author's
imagination or are used fictitiously. Any resemblance to actual
persons, living or dead, business establishments, events, or locales is
entirely coincidental.

I AM NOT A DINOSAUR

Copyright 2020 by Kevin L. Brooks

Illustrations and Artwork by Kevin L. Brooks
Special thanks to Holly Simon for color assistance.

ISBN: 978-0-9960112-3-5

This book was written, illustrated, edited, published and produced in
the United States of America
Made in the USA

I am NOT a DINOSAUR

Note from the author:

As a lifelong lover of nature and science, I strive to portray nature in a positive light. I hope that teaching children about natural history will instill a greater understanding and appreciation of our amazing natural world. So go outside, play, and explore.

We live on a beautiful planet.
This Earth that we call home...

It's filled with amazing and wonderful animals. From the oceans to the mountains and prairies they do roam.

But a long time ago in a time far away,
different animals called this place home..

They too lived from the oceans to the mountains and prairies, but the Earth would shake when they would roam.

This was the age of the
Dinosaurs

Some were fierce...
Some were scary!

Covered with feathers...

Some looked hairy!

Dinosaurs were all shapes and sizes

Some were BIG
and some were SMALL

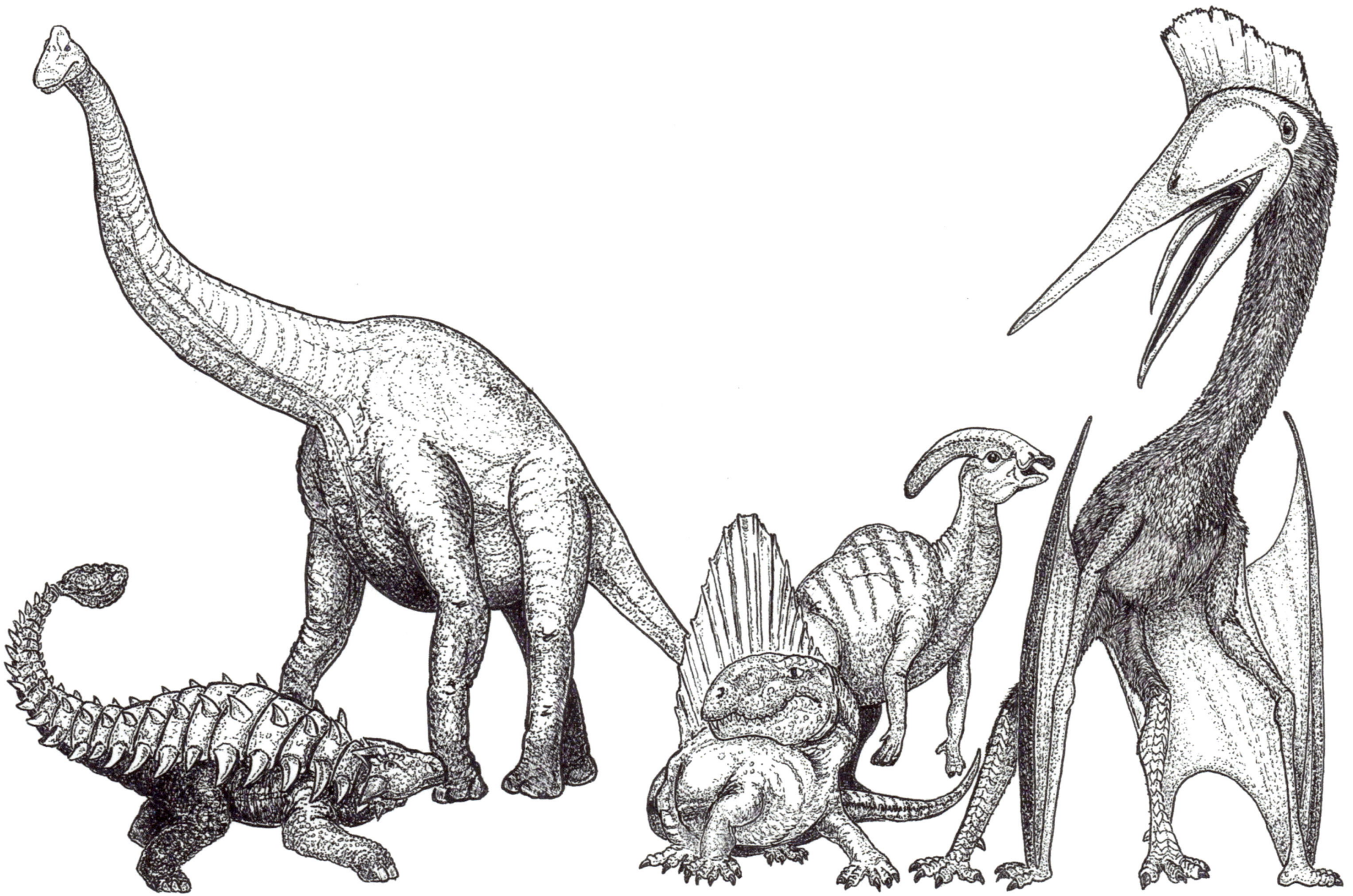

Some were SHORT
and some were TALL

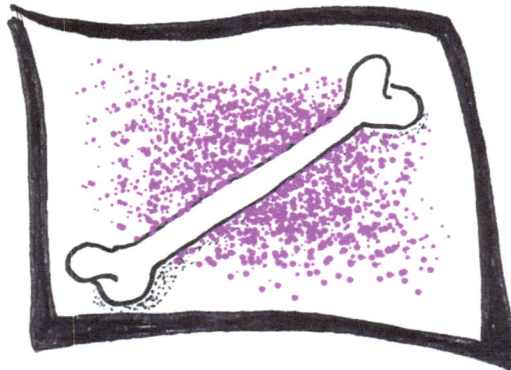

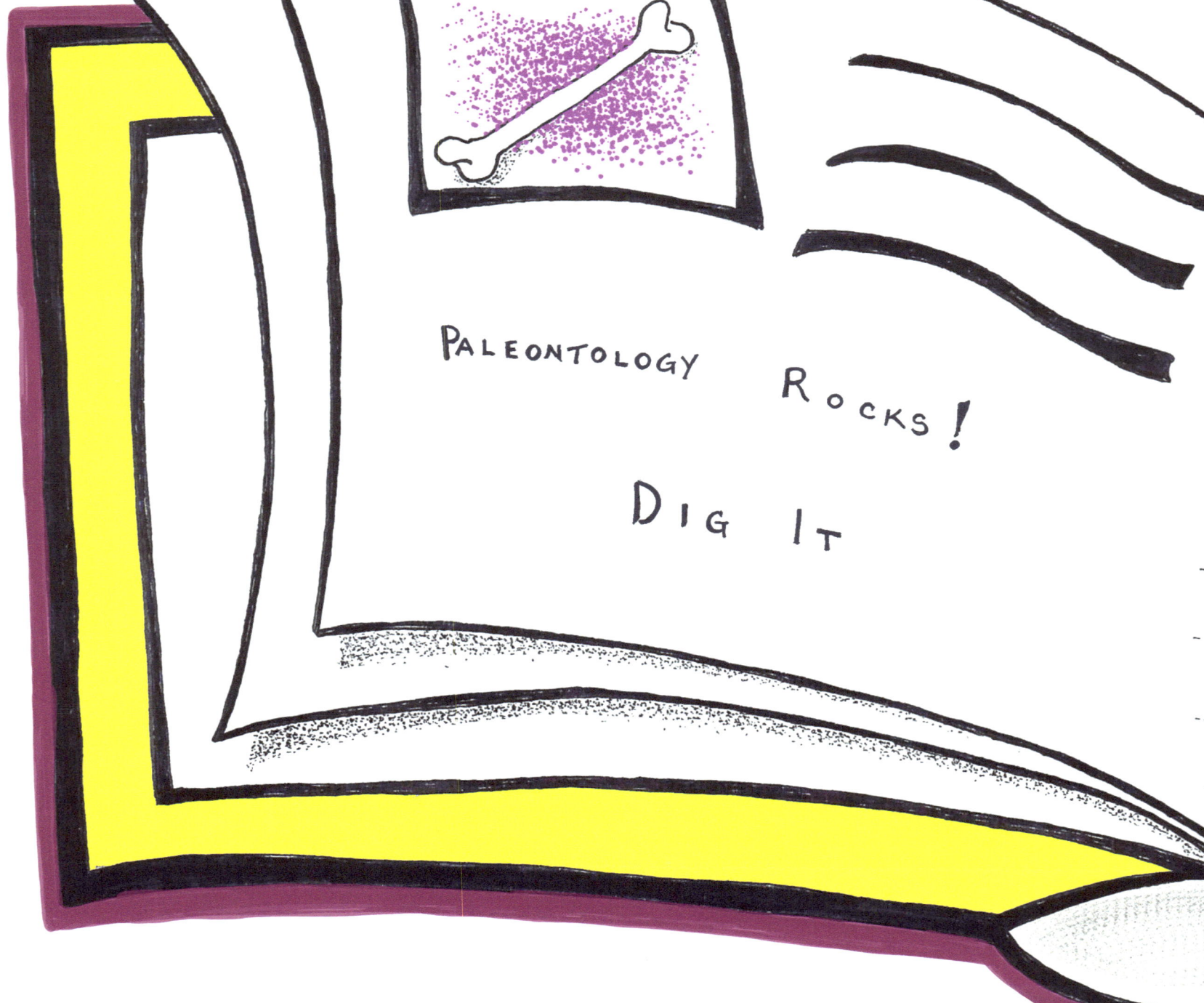

PALEONTOLOGY ROCKS!

DIG It

Hey! Wait a minute! We're not all dinosaurs!

This book is about me!
I'm a Dimetrodon.

...but I am not a dinosaur.

11,700 YEARS AGO

2.6 MYA

5.3 MYA

23 MYA

33 MYA

56 MYA

66 MYA

145 MYA

201 MYA

252 MYA

299 MYA

323 MYA

359 MYA

419 MYA

443 MYA

485 MYA

541 MYA

2.5 BILLION

MYA — MILLIONS OF YEARS AGO

HOLOCEN
PLEISTOCE
PLIOCENE
MIOCEN
OLIGOCEN
EOCENE
PALEOCEN
CRETACEOU
JURASSI
TRIASSIC
PERMIAN
PENNSYLVANIA
MISSISSIPPIA
DEVONIAN
SILURIAN
ORDOVICIA
CAMBRIAN
PROTEROZOI
ARCHEAN
HADEAN

CENOZOIC
MESOZOIC
PALEOZOIC
PRECAMBRIAN

EARTH IS CREATED 4.6 BILLION YEARS AGO

I lived a long time ago, before the dinosaurs during a time called the Permian.

I am not a dinosaur. I'm actually a mammal like reptile called a Synapsid, which means I have a single hole in my skull behind the eye socket, just like you do.

Dinosaurs, birds, and reptiles like lizards, snakes, and crocodiles are Diapsids. They have two holes on each side of their skulls.

The holes are filled with muscles.

Although my bones have been found all over the world, I was originally discovered in the famous Texas Red Beds...

Near the town of Seymour
in Baylor County, Texas.

It is still the best place in the world to study my bones.

In 1878
the famous Paleontologist
Edward Drinker Cope
gave me my name...

Which means
"Two Measures of Teeth."

My teeth show scientists that I was a carnivore. A carnivore eats meat.

...but I am not a dinosaur.

I was also the top predator of my day. A predator is an animal that eats other animals.

...but I am not a dinosaur

Scientists believe I was a good fisherman. The fossil records show that I ate a lot of fish. My favorite was the fresh water shark called Xenacanthus.

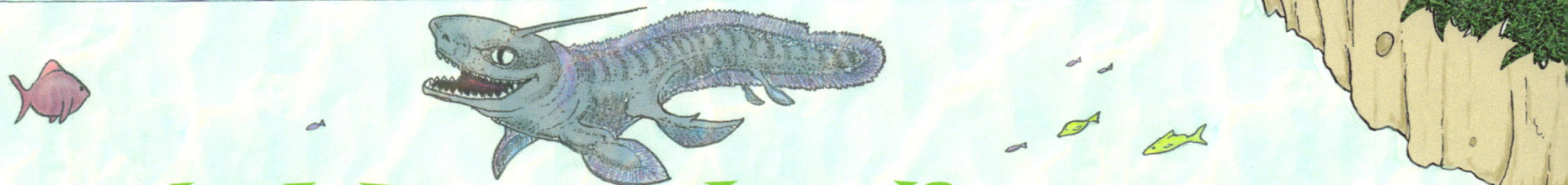

... but I am not a dinosaur.

Dimetrodons had five fingers and five toes.

How many do you have?

How many did dinosaurs have?

Sometimes animals leave an impression in the sand and mud that becomes a fossilized track.

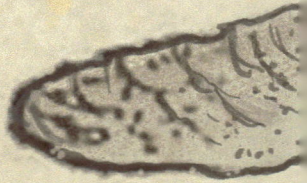

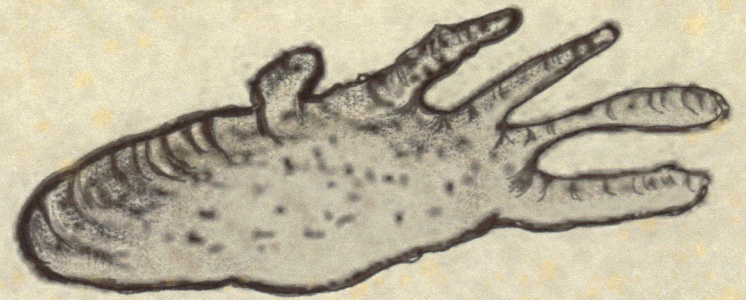

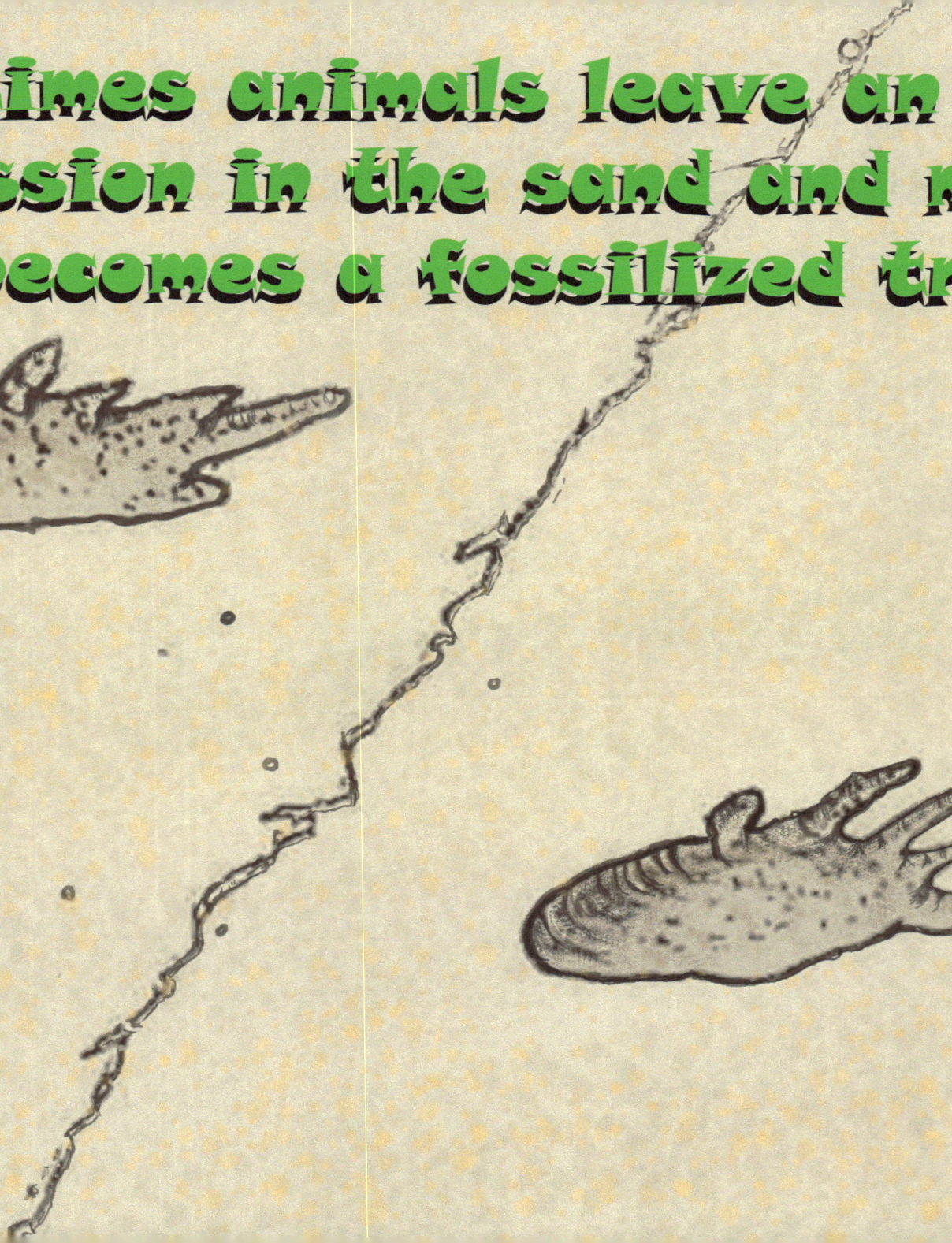

My most prominent feature was the sail on my back. Scientists still argue as to the purpose.

Some people think that my sail was to regulate body temperature. It helped me warm up quickly on cool mornings. This gave me an advantage as a predator because I could hunt while my prey was still groggy.

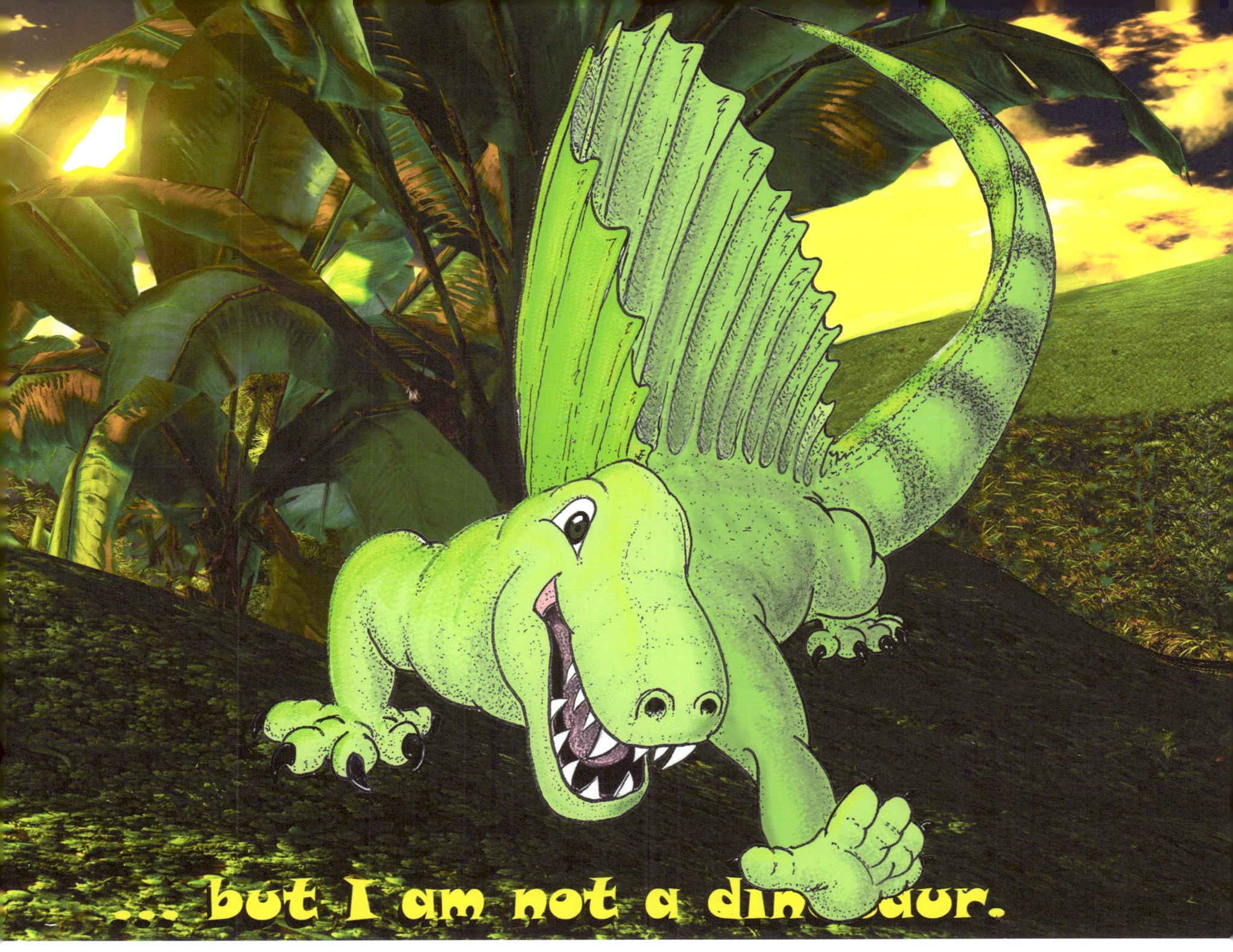

... but I am not a dinosaur.

...but I am not a dinosaur.

My sail also helped me with camoflauge

So I could blend into my surroundings

But most important...
It was to impress the ladies

...but I am not a dinosaur!

Now that you've learned all
about my teeth and my sail,
...that brings us closer
to the end of my tail.

So each night when you lay down your head, look in your closet, look under your bed.

I may be waiting, ready to pounce, but don't be afraid...

I'm lovable. Even if it's just an ounce.

I may look big and scary
but I had no feathers
and I probably wasn't hairy

Long sharp teeth make me look vile
but that's just how I look
when I give you a smile

And if we meet in a long dark alley
Don't scream like a girl
Like Chris and Holly

So, if you find yourself out and about,
Drop by a museum and give me a shout.

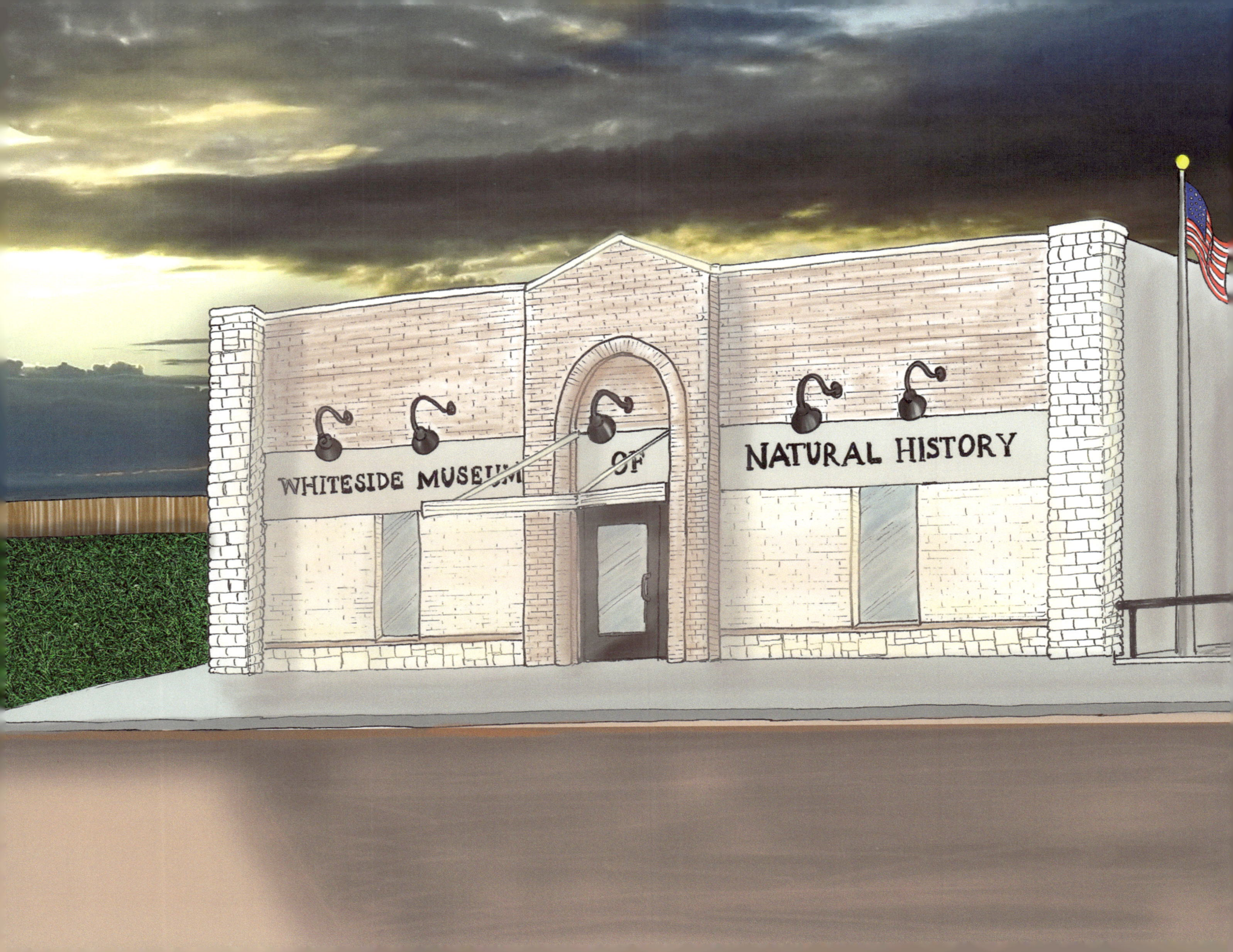

...and always remember
I am not a dinosaur!

I hope you have enjoyed learning about Dimetrodons and the Permian Era. To Learn more, visit a Natural History Museum like the American Museum of Natural History in New York City or the Smithsonian National Museum of Natural History in Washington, DC.

...and for a real treat, visit the Whiteside Museum of Natural History in Seymour, Texas. It is the only museum in the world that specializes in the lower Permian Era. New discoveries about Dimetrodons are made every day.

Follow them at ... www.wmnh.org
or on Facebook
Instagram
Twitter

Dimetrodon is an extinct genus of non-mammalian synapsids, called Pelycosaurs. They are frequently referred to as "mammal-like reptiles." Dimetrodon lived during the Early Permian, known as the Cisuralian Period, around 295-272 million years ago. It is a member of the family Sphenacodontidae. Dimetrodon is often mistaken for a dinosaur, but it became extinct some 40 million years before the first appearance of dinosaurs. Although reptile-like in appearance, Dimetrodon is a synapsid and nevertheless more closely related to mammals than to modern reptiles. Dimetrodon had a tall, curved skull with a single opening behind each eye, known as temporal fenestrae. Dimetrodons had large carnivorous teeth. The size of the teeth varied greatly along the length of the jaws, lending Dimetrodon its name, which means "two measures of teeth". Dimetrodon was the apex predator of its day. There are over a dozen species that have been named since the genus was first described in 1878.

The tale of the sail, the most prominent feature of Dimetrodon is the large sail on its back. This was formed by elongated neural spines extending from the vertebrae. Scientists still argue as to the purpose of this sail. Since all the Early Permian land vertebrates, including Dimetrodon, are assumed to have been cold-blooded, they had to rely on the sun to maintain a high body temperature. The sail of Dimetrodon may have been used to heat and cool its body as a form of thermoregulation. Although this is a contentious matter, some argue that is was a means of warming quickly in the morning sun, and as a way to cool down when body temperature became too high. Others argue that the sail would have been ineffective at removing heat from the body, thus ruling out heat regulation as its main purpose. Others argue that the sail was most likely used in courtship displays for showing off to potential mates or as an intimidation method of threatening rivals.

Here's a funny tale. Dimetrodon's tail makes up a large portion of its total body length. It includes around 50 caudal vertebrae. However, tails were missing or incomplete on the first discovered skeletons. This led many paleontologists in the late nineteenth and early twentieth centuries to believe that Dimetrodon had a very short tail. Early drawings show Dimetrodon with a short stubby tail like a bulldog. It was not until 1927 that a complete tail of Dimetrodon was discovered.

Dimetrodon was originally discovered in Baylor County, Texas in the late 1800's. Most fossils have been found in the southwestern United States, the majority coming from a geological deposit called the Texas Red Beds. Fossils have also been found in Oklahoma, New Mexico, Arizona, Utah and Ohio, as well as Germany.

Recently a new discovery was made concerning a previously known fossil that was collected in 1845 by a farmer digging out a well on Prince Edward Island, Canada. The fossil was a jawbone shown to have steak knife-like teeth. The preeminent paleontologist of the day, Joseph Leidy, identified the fossil as a lower jaw of a dinosaur, similar to the large bipedal species that were being collected in Europe at the time. He named it Bathygnathus borealis. However, recent reexamination of the "dinosaur" fossil has led scientists to change their opinion. It has been renamed to Dimetrodon borealis, making it the first occurrence of a Dimetrodon fossil in Canada.

Fossilized bones of Dimetrodon were first studied during the famous American "Bone Wars" between Edward Drinker Cope and Othniel Charles Marsh. The paleontologist E.D. Cope was the first to study them in the 1870s. Cope obtained the fossils from several collectors who had been exploring a group of rocks in Texas called the Red Beds. Among these fossil hunters were Charles H. Sternberg, the Swiss naturalist Jacob Boll, and the Texas geologist W. F. Cummins. Cope's rival O.C. Marsh also collected some bones of Dimetrodon. Cope named the genus in 1878 in the scientific journal Proceedings of the American Philosophical Society. Most of Cope's specimens went to the American Museum of Natural History or to the University of Chicago's Walker Museum. You can still see them today if you visit the American Museum of Natural History in New York or the Field Museum of Natural History in Chicago.

I am NOT a DINOSAUR

About the author:

Kevin L. Brooks is an avid outdoors enthusiast and naturalist. Growing up in North Texas he lives right smack dab in the middle of the famous Texas Red Beds. This helped influence his love of nature. Although his career revolves around finance, he is a scientist at heart. As a long time board member for the Whiteside Museum of Natural History, Kevin is immersed in the world of Paleontology. This has allowed him to hunt fossils across the greater United States and to meet many of the brightest Paleontologists in the world. And let's face it, he's just a Dino Nerd.

Today he is either standing on a mountain top, exploring a wooded river bottom or digging through some rock and earth in search of some lost and ancient creatures. Or just maybe, he's at home working on his next book.

Kevin is also the author of the bestselling historical fiction series, The Lost Gospel of Barabbas.

www.ingramcontent.com/pod-product-compliance
Lightning Source LLC
Chambersburg PA
CBHW041652260326

41914CB00017B/1616